~A BINGO BOOK~

Physical Science Bingo Book

COMPLETE BINGO GAME IN A BOOK

Written by Rebecca Stark
Educational Books 'n' Bingo

ISBN 978-0-87386-446-6

Educational Books 'n' Bingo

Printed in the U.S.A.

PHYSICAL SCIENCE BINGO
Directions

INCLUDED:

List of Terms

Templates for Additional Terms and Clues

2 Clues per Term

30 Unique Bingo Cards

Markers

1. **Either cut apart the book or make copies of ALL the sheets. You might want to make an extra copy of the clue sheets to use for introduction and review. Keep the sheets in an envelope for easy reuse.**

2. Cut apart the call cards with terms and clues.

3. Pass out one bingo card per student. There are enough for a class of 30.

4. Pass out markers. You may cut apart the markers included in this book or use any other small items of your choice.

5. Decide whether or not you will require the entire card to be filled. Requiring the entire card to be filled provides a better review. However, if you have a short time to fill, you may prefer to have them do the just the border or some other format. Tell the class before you begin what is required.

6. There are 50 topics. Read the list before you begin. If there are any topics that have not been covered in class, you may want to read to the students the topic and clues before you begin.

7. There is a blank space in the middle of each card. You can instruct the students to use it as a free space or you can write in answers to cover topics not included. Of course, in this case you would create your own clues. (Templates provided.)

8. Shuffle the cards and place them in a pile. Two or three clues are provided for each topic. If you plan to play the game with the same group more than once, you might want to choose a different clue for each game. If not, you may choose to use more than one clue.

9. Be sure to keep the cards you have used for the present game in a separate pile. When a student calls, "Bingo," he or she will have to verify that the correct answers are on his or her card AND that the markers were placed in response to the proper questions. Pull out the cards that are on the student's card keeping them in the order they were used in the game. Read each clue as it was given and ask the student to identify the correct answer from his or her card.

10. If the student has the correct answers on the card AND has shown that they were marked in response to the *correct questions,* then that student is the winner and the game is over. If the student does not have the correct answers on the card OR he or she marked the answers in response to *the wrong questions,* then the game continues until there is a proper winner.

11. If you want to play again, reshuffle the cards and begin again.

Have Fun!

TERMS

ACOUSTICS

AERODYNAMICS

ARCHIMEDES

ATOM

BUOYANCY

CHEMISTRY

COLOR

COMPOUND MACHINES

CONDUCTION OF HEAT

CONDUCTOR

CONSERVATION OF ENERGY

CONVECTION

DENSITY

ELECTRICITY

ENERGY

MICHAEL FARADAY

FORCE

FRICTION

FUSION

GEARS

GENERATOR

GRAVITY

HEAT ENERGY

INCLINED PLANE

INERTIA

INSULATOR

LEVER

LIGHT ENERGY

MAGNET

MASS

MATTER

MOLECULE

MOTION

SIR ISAAC NEWTON

NUCLEUS

OPTICS

PHYSICS

PRISM

PULLEY

REFLECTION

REFRACTION

SCREW

SIMPLE MACHINES

SOUND

SPECTRUM

TEMPERATURE

THEORY

WAVES

WEDGE

WHEEL AND AXLE

Additional Terms

Choose as many additional terms as you would like and write them in the squares. Repeat each as desired.
Cut out the squares and randomly distribute them to the class.
Instruct the students to place their square on the center space of their card.

Clues for Additional Terms

Write three clues for each of your additional terms.

_____ 1. 2. 3.	_____ 1. 2. 3.
_____ 1. 2. 3.	_____ 1. 2. 3.
_____ 1. 2. 3.	_____ 1. 2. 3.

© Barbara M. Peller

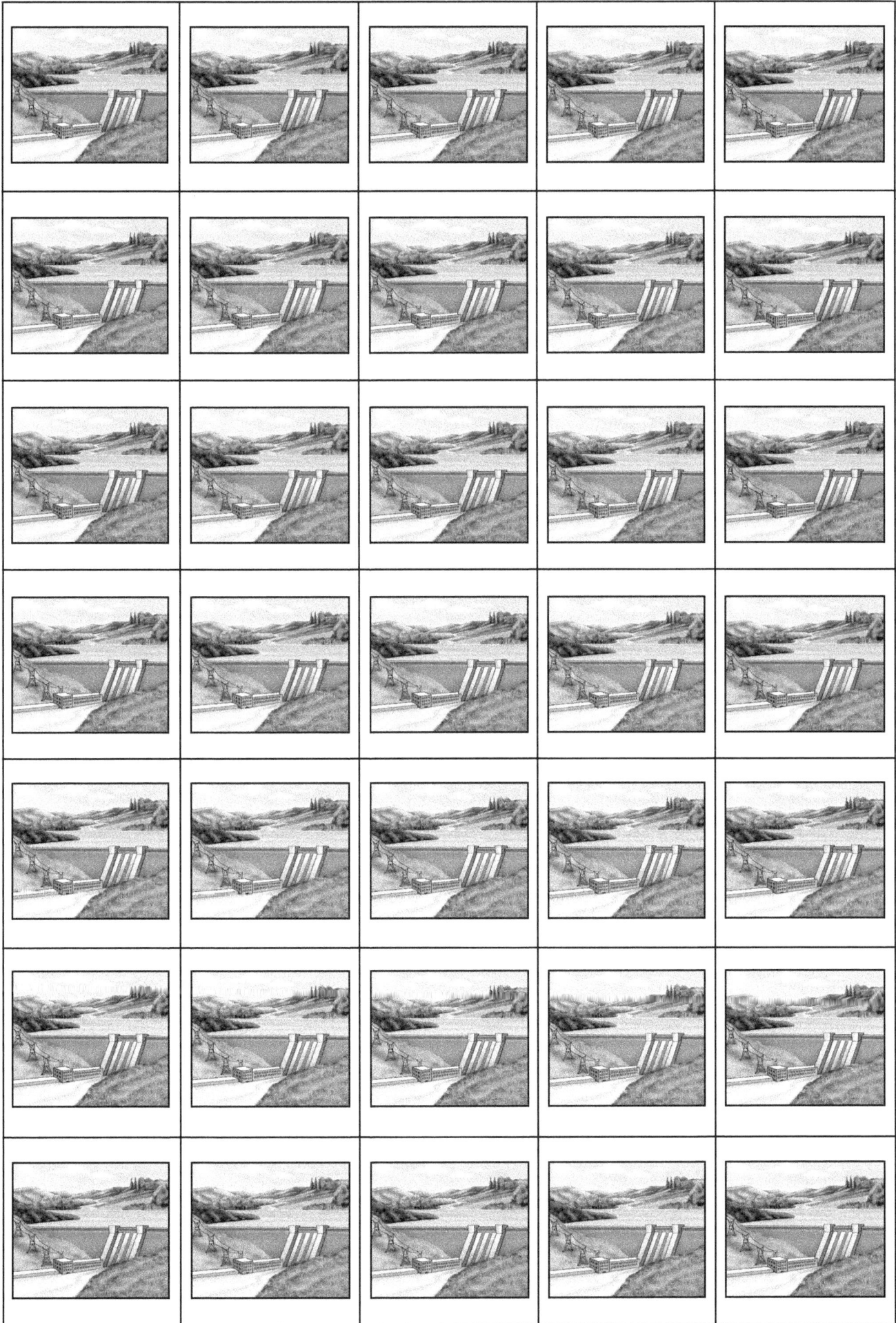

ACOUSTICS 1. This science deals with the study of sound. 2. The term comes from an ancient Greek word meaning "able to be heard." 3. This field of study encompasses ultrasound and infrasound as well as sound audible to the human ear.	**AERODYNAMICS** 1. This is the study of the motion of air, especially when it interacts with a moving object. 2. The word comes from two Greek words: *aerios,* meaning "concerning the air," and *dynamis*, meaning "force." 3. Vortices are one of the phenomena associated with this branch of classical mechanics.
ARCHIMEDES 1 The principle named for him states that a body immersed in a fluid is buoyed up by a force equal to the weight of the displaced fluid. 2. This Greek scientist, who lived from c. 287 to 212 BCE, is best known for the principle explaining how things float. 3. This Greek scientist explained how pressure changes with depth in liquids and gases.	**ATOM** 1. It is the smallest component of an element. 2. Protons, which carry a positive charge, and neutrons, which carry no charge, make up its center, or nucleus. 3. Electrons, which have a positive charge, circle the nucleus of an ___.
BUOYANCY 1. It is an object's ability to float. 2. It is the upward force on an object produced by the liquid or gas in which it is fully or partially immersed. 3. Archimedes' Principle has to do with this concept.	**CHEMISTRY** 1. This branch of science deals with the composition of substances and their properties and reactions. 2. The study of acids and bases is included in this branch of science. 3. The Periodic Table is used in this branch of science.
COLOR 1. The wavelength of light determines the perceived ___. 2. Red, green and blue are called primary ___s of light. 3. The primary ___s in painting are magenta, yellow and cyan.	**COMPOUND MACHINES** 1. These are made up of two or more simple ones. 2. A bicycle is one. It is made up of screws, wheels and axles, levers and pulleys working together. 3. A pair of scissors is one. It uses levers to force the wedges, in the form of blades, to cut.
CONDUCTION OF HEAT 1. It is the transfer or heat from warm areas to cooler areas. 2. If you stir a hot drink with a metal spoon, the spoon gets hot because of this. 3. Fourier's Law, which deals with this, states that if there is a temperature gradient within a body, heat energy will flow from the region of high temperature to the region of low temperature.	**CONDUCTOR** 1. It is a material that allows electricity or heat to flow through it easily. 2. Metals are good ones. Water is a good ___ of electricity. 3. An antonym is "insulator."

Physical Science Bingo

© Barbara M. Peller

CONSERVATION OF ENERGY

1. This law states that the total amount of energy in a system remains the same.
2. This law states that although the amount of energy remains constant, it may change forms.
3. According to this law, energy can be neither created nor destroyed.

CONVECTION

1. This is the transfer of heat energy in a gas or liquid by movement of currents.
2. An example of this process is heat leaving a cup of coffee or tea as the currents of steam and air rise. The heat moves with the steam.
3. Heating a pot of water on a stove is an example of the transfer of heat by this process.

DENSITY

1. This is a measure of how tightly the matter within a mass is packed together.
2. It can be thought of as how much something weighs in relation to its size.
3. This physical property of matter describes the degree of compactness of a substance—in other words, how closely packed together the atoms of an element or molecules of a compound are.

ELECTRICITY

1. Current __ is the flow of electrons in closed loops, or circuits. It must have a complete path before the electrons can move.
2. ___ is a secondary energy source; it results from the conversion of other sources of energy, such as coal, natural gas, or oil.
3. This energy source is measured in units of power called watts.

ENERGY

1. There are many forms of this, but they can all be put into two categories: kinetic and potential.
2. Some forms include heat, light, sound, electrical, nuclear and motion.
3. Sources may be categorized as renewable, such as wind, and nonrenewable, such as oil.

MICHAEL FARADAY

1. This 19th-century English scientist discovered electromagnetic induction.
2. This English scientist invented the electric transformer, the electric motor and the dynamo.
3. His discovery that a suspended magnet revolved around a current-bearing wire led him to theorize that magnetism is a circular force.

FORCE

1. It is the push or pull that makes things move, change shape, or change because of an object's interaction with another object.
2. It is measured using the standard metric unit known as the newton (N).
3. They come in pairs called actions and reactions. For every action, there is an equal and opposite reaction.

FRICTION

1. This force resists relative motion between two bodies in contact.
2. This force stops stationary things from moving and slows down things already in motion.
3. Lubricants are used to reduce this force.

FUSION

1. This the union of the nuclei of different atoms to form heavier nuclei resulting in enormous quantities of nuclear energy.
2. An antonym of this is "fission," or the splitting of a nucleus.
3. Nuclear energy can result from either fission or ___.

GEARS

1. These toothed wheels are a type of simple machine.
2. These work by one wheel turning the next.
3. In these simple machines, the drive wheel can make a larger wheel turn more slowly or a smaller wheel turn more quickly.

Physical Science Bingo

© Barbara M. Peller

GENERATOR 1. An electrical one is a machine that converts mechanical energy into electrical energy. 2. It usually uses electromagnetic induction to convert mechanical energy into electrical energy. 3. A dynamo is this kind of compound machine.	**GRAVITY** 1. It is the natural force of attraction between any two objects with mass. 2. It is what keeps our moon in Earth's orbit and the planets in orbit around the sun. 3. Newton's Law of ___ states that every particle in the universe attracts every other particle.
HEAT ENERGY 1. This form of energy moves, spreading out from hotter things to cooler things. The energy is created by the motion of the atoms and molecules. 2. A unit of this kind of energy is called a joule. 3. This type of energy moves by conduction, convection and radiation.	**INCLINED PLANE** 1. This simple machine is a flat surface whose endpoints are at different heights. 2. A ramp is an example of this kind of simple machine. 3. Garbage chutes that are often found in high-rise apartment buildings are examples of this type of simple machine.
INERTIA 1. This is the tendency of objects to resist a change in their state of motion. This principle is described in Newton's First Law of Motion. 2. It is the principle that an object at rest tends to stay at rest. 3. It is the principle that an object in motion tends to stay in motion with the same speed and in the same direction unless acted upon by an unbalanced force.	**INSULATOR** 1. It is a material that resists the flow of electric current or heat. 2. It is the opposite of "conductor." 3. Glass, rubber and plastic are often used as an ___ of electricity.
LEVER 1. This kind of simple machine involves moving a load around a pivot using a force. 2. An example of this kind of simple machine is a hammer when used to *remove* nails. 3. There are three classes of this simple machine. They are classified according to the location of the fulcrum.	**LIGHT** 1. It is the only form of energy that we can see directly. 2. Opacity, translucence and transparency have to do with how much of this passes through something. 3. Green plants convert this type of energy into chemical energy through photosynthesis.
MAGNET 1. It is a material that produces a force which attracts iron, nickel or cobalt. 2. Lodestone is a natural one. 3. If one is held freely, it would line up with Earth's north and south poles.	**MASS** 1. This is the property that causes matter to have weight in a gravitational field. On Earth's surface it is the same as weight. 2. It is the amount of matter in something. 3. The SI unit of this is the kilogram (kg). (Note: SI = International System of Units)

Physical Science Bingo

MATTER	**MOLECULE**
1. It occurs in three main states, or phases: solid, liquid and gas. 2. Its fourth state, plasma, occurs when electrons are stripped away by high heat or pressure. 3. It changes from a solid state to a liquid state at its melting point.	1. It is the simplest structural unit of an element or a compound and is made up of even smaller particles called atoms. 2. It is a unit of atoms bonded together. Some are compounds. 3. If it contains at least two different elements, then it is a compound.
MOTION	**SIR ISAAC NEWTON**
1. It is a continuous change in the location of a body as the result of applied force. 2. Newton's First Law of ___ is sometimes referred to as the Law of Inertia. 3. Newton's First Law states that an object in ___ tends to remain in this state unless an external force is applied to it.	1. His First Law of Motion deals with the concept of inertia. 2. His Second Law of Motion states that the acceleration of an object is dependent upon the net force acting upon the object and the mass of the object. 3. His Third Law of Motion states that for every action there is an equal and opposite reaction.
NUCLEUS	**OPTICS**
1. Every atom has one; it contains protons and neutrons. 2. This part of an atom contains most of its mass. 3. Tiny electrons move around it.	1. It is the branch of physical science that studies the physical properties of light. 2. Among the properties of light studied in this branch of science are intensity, frequency and polarization. 3. This branch of science also studies the interaction of light with matter.
PHYSICS	**PRISM**
1. This is the science of matter and energy and their interactions. 2. Fields within this science include acoustics, optics, mechanics, and electromagnetism. 3. Motion, force and simple machines are concepts in this science.	1. This is a transparent optical element with flat, polished surfaces that refract light. 2. It can be used to break light up into the colors of the spectrum. 3. Rainbows form because each drop of rain acts like a ___ and separates the light into its constituent colors.
PULLEY	**REFLECTION**
1. This simple machine consists of a wheel with a groove in which a rope, cable, chain or belt can run to change the direction of the pull. 2. Adding more than one wheel to this simple machine makes it possible to lift heavier loads. 3. A block & tackle is a compound machine made of two or more of simple machine with a rope or cable threaded between them.	1. It is the return of light, sound or other waves from a surface. 2. Light ___ is called specular if the image is retained. A mirror is the most common example of this. 3. Diffuse ___ occurs when light strikes a rough or granular surface, causing it to bounce off in all directions.

Physical Science Bingo

REFRACTION
1. This phenomenon occurs because as light passes from one transparent medium to another, it changes speed and bends.
2. If you submerge a straight stick into water, the stick will appear bent at the point where it enters the water. This is caused by ___.
3. Lenses work because of this phenomenon.

SCREW
1. This simple machine is actually an inclined plane that is shaped like a helix.
2. It can be described as an inclined plane wrapped around a shaft.
3. The width of its thread is like the angle of an inclined plane. The wider the thread, the harder it is to turn it.

SIMPLE MACHINES
1. They include the lever, the inclined plane, the wheel and axle, the wedge, the screw, and the pulley.
2. If two or more are combined, the result is a compound machine.
3. Machines are devices that make work easier to perform. All are built from one or more of these.

SOUND
1. The number of vibrations per second is the frequency of the ___.
2. This type of energy travels more slowly than light energy.
3. The more quickly something vibrates, the higher the ___.

SPECTRUM
1. It is the distribution of colors produced when white light is separated into its constituent colors.
2. Colors of the visible ___ include violet, indigo, blue, green, yellow, orange, and red.
3. A prism can be used to separate white light into the colors of the ___.

TEMPERATURE
1. As used in the physical sciences, this is a measurement of the average kinetic energy in a sample—in other words, how fast the molecules are vibrating.
2. Three scales used to measure this value are Kelvin, Celsius, and Fahrenheit.
3. We commonly refer to it as the degree of hotness or coldness of a body or environment.

THEORY
1. A scientifically acceptable general principle or body of principles explaining a phenomenon is often called a ___.
2. It is an explanation or model based on observation, experimentation, and reasoning, especially one that has been tested.
3. Albert Einstein is known for his ___ of Relativity.

WAVES
1. They are disturbances or variations which travel through a medium.
2. They travel and transfer energy from one point to another. Examples are seismic ___, sound ___, radio ___ and light ___.
3. Periodic ones are characterized by crests, or highs, and troughs, or lows.

WEDGE
1. This simple machine is actually two back-to-back inclined planes. It is used to separate, to hold and to tighten.
2. An axe is an example of this simple machine that is used to split, or separate.
3. A doorstop is an example of this simple machine that is used to tighten.

WHEEL AND AXLE
1. This simple machine is a actually a lever that rotates in a circle around a center point, or fulcrum.
2. Bicycle wheels, ferris wheels and gears are examples.
3. This simple machine consists of a large wheel rigidly secured to a smaller wheel or shaft, called an axle.

Physical Science Bingo

Simple Machines	Acoustics	Buoyancy	Inertia	Michael Faraday
Conservation of Energy	Aerodynamics	Theory	Motion	Sir Isaac Newton
Atom	Wedge		Optics	Chemistry
Wheel and Axle	Force	Temperature	Mass	Nucleus
Prism	Gears	Inclined Plane	Waves	Light Energy

Physical Science Bingo

Wheel and Axle	Atom	Magnet	Sound	Heat Energy
Nucleus	Electricity	Compound Machines	Force	Matter
Convection	Gears		Friction	Temperature
Reflection	Screw	Wedge	Refraction	Light Energy
Sir Isaac Newton	Theory	Inclined Plane	Conservation of Energy	Waves

Physical Science Bingo

Wheel and Axle	Temperature	Electricity	Mass	Atom
Motion	Aerodynamics	Conductor	Acoustics	Insulator
Force	Theory		Matter	Archimedes
Wedge	Convection	Prism	Reflection	Magnet
Waves	Conservation of Energy	Inclined Plane	Refraction	Heat Energy

Physical Science Bingo

Wedge	Matter	Buoyancy	Conservation of Energy	Heat Energy
Lever	Color	Acoustics	Sound	Atom
Optics	Reflection		Michael Faraday	Inertia
Temperature	Fusion	Theory	Inclined Plane	Compound Machines
Conduction of Heat	Sir Isaac Newton	Physics	Waves	Chemistry

Physical Science Bingo

Sir Isaac Newton	Michael Faraday	Force	Compound Machines	Conservation of Energy
Lever	Temperature	Conductor	Friction	Aerodynamics
Buoyancy	Chemistry		Motion	Generator
Light Energy	Heat Energy	Simple Machines	Refraction	Density
Electricity	Inclined Plane	Atom	Wedge	Optics

Physical Science Bingo

Archimedes	Matter	Magnet	Heat Energy	Chemistry
Mass	Force	Density	Acoustics	Atom
Sound	Conduction of Heat		Color	Friction
Inclined Plane	Prism	Refraction	Physics	Buoyancy
Nucleus	Compound Machines	Simple Machines	Optics	Fusion

Physical Science Bingo

Simple Machines	Matter	Generator	Motion	Electricity
Nucleus	Heat Energy	Gears	Aerodynamics	Lever
Magnet	Inertia		Friction	Color
Wedge	Reflection	Conductor	Wheel and Axle	Convection
Inclined Plane	Conservation of Energy	Refraction	Physics	Archimedes

Physical Science Bingo

Optics	Matter	Energy	Mass	Color
Lever	Buoyancy	Sound	Chemistry	Compound Machines
Fusion	Pulley		Heat Energy	Michael Faraday
Waves	Wedge	Wheel and Axle	Conduction of Heat	Reflection
Theory	Inclined Plane	Physics	Force	Nucleus

© Barbara M. Peller

Physical Science Bingo

Friction	Electricity	Gears	Fusion	Conservation of Energy
Conduction of Heat	Heat Energy	Optics	Force	Matter
Insulator	Simple Machines		Aerodynamics	Energy
Density	Light Energy	Prism	Motion	Generator
Reflection	Refraction	Conductor	Wheel and Axle	Michael Faraday

Physical Science Bingo

Wheel and Axle	Mass	Color	Sound	Fusion
Chemistry	Compound Machines	Acoustics	Aerodynamics	Heat Energy
Pulley	Matter		Inertia	Convection
Prism	Light Energy	Density	Refraction	Insulator
Conductor	Nucleus	Magnet	Sir Isaac Newton	Optics

Physical Science Bingo

Archimedes	Matter	Force	Density	Nucleus
Energy	Insulator	Motion	Friction	Acoustics
Lever	Heat Energy		Magnet	Gears
Conductor	Atom	Refraction	Conservation of Energy	Wheel and Axle
Conduction of Heat	Inclined Plane	Simple Machines	Physics	Electricity

Physical Science Bingo

Electricity	Michael Faraday	Insulator	Mass	Friction
Gears	Nucleus	Buoyancy	Physics	Aerodynamics
Simple Machines	Generator		Chemistry	Sound
Inclined Plane	Reflection	Heat Energy	Wheel and Axle	Lever
Matter	Energy	Pulley	Conduction of Heat	Compound Machines

Physical Science Bingo

Density	Michael Faraday	Archimedes	Insulator	Chemistry
Buoyancy	Energy	Heat Energy	Friction	Convection
Mass	Compound Machines		Gears	Generator
Optics	Refraction	Color	Pulley	Wheel and Axle
Inclined Plane	Light Energy	Physics	Simple Machines	Motion

© Barbara M. Peller

Physical Science Bingo

Conservation of Energy	Heat Energy	Force	Friction	Conduction of Heat
Compound Machines	Simple Machines	Insulator	Aerodynamics	Matter
Density	Inertia		Magnet	Conductor
Light Energy	Refraction	Pulley	Color	Archimedes
Inclined Plane	Sound	Convection	Nucleus	Optics

Physical Science Bingo

Motion	Friction	Force	Electricity	Mass
Archimedes	Magnet	Acoustics	Buoyancy	Conduction of Heat
Chemistry	Simple Machines		Atom	Matter
Inclined Plane	Insulator	Energy	Refraction	Density
Nucleus	Reflection	Physics	Fusion	Gears

Physical Science Bingo

Color	Insulator	Energy	Fusion	Screw
Sound	Convection	Generator	Lever	Inertia
Density	Michael Faraday		Chemistry	Gears
Wedge	Compound Machines	Inclined Plane	Motion	Wheel and Axle
Conduction of Heat	Spectrum	Physics	Reflection	Matter

Physical Science Bingo

Conductor	Molecule	Gravity	Insulator	Conservation of Energy
Motion	Conduction of Heat	Refraction	Inertia	Generator
Friction	Optics		Spectrum	Energy
Light Energy	Nucleus	Wheel and Axle	Force	Convection
Prism	Density	Electricity	Mass	Michael Faraday

Physical Science Bingo

Fusion	Pulley	Compound Machines	Density	Sound
Matter	Conductor	Prism	Chemistry	Conduction of Heat
Friction	Convection		Gravity	Buoyancy
Light Energy	Acoustics	Refraction	Wheel and Axle	Magnet
Spectrum	Insulator	Force	Molecule	Archimedes

Physical Science Bingo

Chemistry	Archimedes	Insulator	Energy	Pulley
Motion	Mass	Matter	Electricity	Inertia
Molecule	Conservation of Energy		Aerodynamics	Atom
Magnet	Spcotrum	Prism	Reflection	Gravity
Buoyancy	Screw	Nucleus	Optics	Physics

© Barbara M. Peller

Physical Science Bingo

Pulley	Molecule	Mass	Insulator	Physics
Compound Machines	Gears	Lever	Prism	Sound
Michael Faraday	Generator		Wedge	Acoustics
Sir Isaac Newton	Theory	Waves	Reflection	Spectrum
Temperature	Optics	Screw	Wheel and Axle	Gravity

Physical Science Bingo: Card No. 20

Physical Science Bingo

Motion	Archimedes	Lever	Insulator	Sir Isaac Newton
Michael Faraday	Gravity	Color	Energy	Simple Machines
Convection	Nucleus		Molecule	Force
Prism	Electricity	Spectrum	Light Energy	Optics
Wedge	Screw	Physics	Conductor	Reflection

Physical Science Bingo

Fusion	Magnet	Gravity	Buoyancy	Density
Sound	Mass	Atom	Energy	Aerodynamics
Compound Machines	Inertia		Simple Machines	Generator
Spectrum	Light Energy	Reflection	Acoustics	Conservation of Energy
Screw	Conductor	Molecule	Convection	Lever

Physical Science Bingo

Color	Molecule	Electricity	Buoyancy	Physics
Archimedes	Pulley	Nucleus	Motion	Acoustics
Magnet	Density		Waves	Simple Machines
Convection	Screw	Spectrum	Conductor	Reflection
Sir Isaac Newton	Theory	Optics	Prism	Gravity

Physical Science Bingo

Color	Pulley	Conservation of Energy	Molecule	Energy
Gravity	Physics	Lever	Sound	Simple Machines
Generator	Fusion		Density	Convection
Sir Isaac Newton	Waves	Spectrum	Conductor	Michael Faraday
Temperature	Wedge	Screw	Mass	Theory

Physical Science Bingo: Card No. 24

Physical Science Bingo

Wedge	Lever	Molecule	Force	Gravity
Acoustics	Light Energy	Motion	Color	Aerodynamics
Michael Faraday	Energy		Waves	Spectrum
Atom	Sir Isaac Newton	Theory	Screw	Inertia
Physics	Conservation of Energy	Compound Machines	Conduction of Heat	Temperature

Physical Science Bingo

Gravity	Molecule	Magnet	Sound	Fusion
Prism	Mass	Energy	Pulley	Color
Light Energy	Waves		Inertia	Wedge
Conductor	Buoyancy	Sir Isaac Newton	Screw	Spectrum
Generator	Conduction of Heat	Force	Theory	Temperature

Physical Science Bingo

Magnet	Compound Machines	Molecule	Pulley	Gears
Sir Isaac Newton	Waves	Motion	Spectrum	Aerodynamics
Refraction	Theory		Screw	Wedge
Fusion	Archimedes	Lever	Temperature	Acoustics
Conduction of Heat	Inertia	Gravity	Atom	Generator

Physical Science Bingo

Chemistry	Pulley	Atom	Molecule	Color
Gears	Gravity	Waves	Sound	Inertia
Theory	Convection		Generator	Prism
Wheel and Axle	Fusion	Nucleus	Screw	Spectrum
Buoyancy	Friction	Conduction of Heat	Temperature	Sir Isaac Newton

Physical Science Bingo

Gravity	Pulley	Fusion	Motion	Friction
Light Energy	Prism	Lever	Generator	Atom
Michael Faraday	Waves		Aerodynamics	Molecule
Gears	Sir Isaac Newton	Hoat Energy	Screw	Spectrum
Color	Energy	Temperature	Archimedes	Theory

Physical Science Bingo

Conservation of Energy	Molecule	Sound	Friction	Spectrum
Acoustics	Pulley	Magnet	Inertia	Aerodynamics
Light Energy	Density		Generator	Lever
Temperature	Archimedes	Buoyancy	Screw	Waves
Sir Isaac Newton	Electricity	Theory	Gravity	Atom